POETS	ILLUSTRATORS	ECOLOGICAL ADVISORS
Angela A. Coyer	Angela A. Coyer	Daniel J. Kane
Lora A. Vieman	Judith A. Kane	Diane M. Kane
Rebecca J. Polum	Rebecca J. Polum	
Teresa M. Veraguth	Teresa M. Veraguth	

Dad always said... Many hands make light work

***This book has been created by incredible volunteers who believe we all can make a difference.
Many hours have been labored in love for our beautiful planet. Thank-you, thank-you, thank-you, to all of you!!***

Teresa's family and friends helped create this book, together, with her. Her mom, Judy, a beautiful artist in her own right, gave it depth by detailing some of the watercolor backgrounds. Teresa's niece, Rebecca, also gracefully swirled out backgrounds, as well as wrote one of the poems. Teresa's brother and his wife, Dan & Diane, both with a lifetime of experience and education in ecology, biology, geology and conservation, helped advise for accurate ecologic and conservation content. Her college roommate and lifelong friend, Lora, made the chapters come to life with her vivid poetry, as well as helped "massage" other wording where it was needed. Friend and fellow graphic designer, Angela, likes to tell people she helped "color". Her beautiful art lights up many of the pages.

In addition, our team would like to say a heartfelt thank you to the following...
A special group of students at the school where Teresa substitutes (you know who you are) who listened and cheered on our work!
Our gratitude goes out to, Teresa Sullivan & Amy Wolkowinsky, two beautiful ladies who magnanimously supported this endeavor.
Without the unwavering grace, patience, love and support of Teresa's husband, Pat Veraguth, this book would not exist.
Last but not least, our team, would like to honor and thank our creator for the inspiration and guidance for this project.

Artist & Editorial Credits

Lora Wieman: "Where Do I Roam?" poem, Blanchard's Cricket frog poem, Dakota Skipper butterfly poem, Red-headed Woodpecker poem.
Angela Coyer: Piping Plover bird art, light house and contributed to the poem, Dakota Skipper butterfly art, street lamp light and contributed to poem.
Rebecca Polum: Watercolor background art for the Blanchard's Cricket frog, Dakota Skipper butterfly, Piping Plover, ZigZag Darner dragonfly and the Western moose. In addition, Rebecca wrote the poem for the dragonfly.
Judith Kane: Watercolor background art for Red-Headed woodpecker, Walleye and Common loon.
Teresa Veraguth: Art for the Blanchard's Cricket frog, the Red-headed woodpecker, Walleye, ZigZag Darner Dragonfly, Loon and Western moose. Art for moose lantern and birch, poems for Walleye, Common loon, Western moose and contributed to the Piping Plover poem. Also responsible for research, graphic design layout, green holiday calendar and final editing.
Daniel & Diane Kane: Many hours spent advising for accurate ecologic and conservation content.

THIS BOOK IS DEDICATED TO PLANET
EARTH

To: Connor,
Get outdoors and find some grand adventures!
Hopefully these pages "shine a light" on fun things you can do!"
- TV

This book was printed and published in the state of Minnesota, USA, using sustainable practices and materials.
No part of this publication may be reproduced, stored in a retrieval system, or transmitted in any form or by any means, electronic, mechanical, photocopied, recorded, or otherwise without written permission of the publisher.
Write to Froglegs & Co, 953 90th St SW, Alexandria, MN, 56308

Copyright © 2024 Froglegs & Co
All Rights Reserved by Teresa Veraguth and

Froglegs & Co

First Edition of the First Book in the Small Hands Can Help series.
1/ 51
ISBN: 979-8-9910451-0-0

Contents

*Our mission is to empower and engage families,
educators, and communities to take actions that support nature and us.
When we all join together, little things small can hands do will make a big difference!*

Inside front cover... *about the author* - Author Biography

Page 5.......... *why we did this* - About this book

Page 6.......... *follow the wind and stars* - Preface: How to navigate this book

Page 7.......... *hop to the Acris blanchardi* - Chapter 1 - Blanchard's Cricket Frog

Page 11.......... *flutter by the Hesperia dacotae* - Chapter 2 - Dakota Skipper Butterfly

Page 15.......... *swoop in on the Melanerpes erythrocephalus* - Chapter 3 - Red-headed Woodpecker

Page 19.......... *swim with the Sander vitreus* - Chapter 4 - Walleye

Page 23.......... *march up to the Charadrius melodus* - Chapter 5 - Piping Plover

Page 27.......... *zip to the Aeshna sitchensis* - Chapter 6 - Zigzag Darner Dragonfly

Page 31.......... *dive with the Gavia immer* - Chapter 7 - Common Loon

Page 35.......... *amble over to the Alces alces* - Chapter 8 - Western Moose

Page 40.......... *fill up your days with eco-friendly fun* - Environmental Green Holiday Calendar

Page 42.......... *what do YOU think* - Sparks of Thought

Page 43.......... *give credit where credit is due* - References

Inside back cover... *puzzle it out* - Vocabulary Crossword Puzzle

 # ABOUT THIS BOOK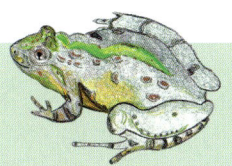

The **Small Hands Can Help** book series grew out of conversations and observations about ways to bring families outdoors and foster a love of all things Earth friendly. In this book, parents, grandparents and educators can learn delightful ways to connect with their local habitats. The book shows how they are part of their ecosystem and shines a light on programs, technologies and volunteer opportunities that can be participated in, using free, easy, effective, and trusted resources.

Families can learn bird names and begin to identify birds from their songs. Simple and safe apps will show them the names of the insects, and plants or mushrooms they find in their own backyards or parks, without collecting personal data. These chapters present poems filled with vivid scientific facts about species that are disappearing from the beautiful and diverse state of **Minnesota**. Since this is the first book in a book series that will include a book for each state, plus Washington D.C. families will know they can learn about other states and use them in their travels too.

This series is designed for the readers return to the pages over and over, to use it differently at different ages. For example, a toddler may learn to read and recite the *Shine a Light* pages or the *Roam* poem, while a tween-ager might use a fact filled species poem to write a short essay for school or find an activity to use for the fair or for a 4-H project.

The citizen science apps and other quality resources can help teens and young adults create or join projects that can bring friends together in actions of important change. A high school student may want to conduct a frog call survey, as found in the first chapter, for an FFA, or science project. A grandparent may want to sign up for the iNaturalist app, linked via a QR code in chapter two, to help them identify the various organisms they encounter during their walks. The entire family may decide on September 1st to observe World Beach Day, (check out the Environmental Green Holiday calendar in the back of the book). After all, Minnesota is the Land of 10,000 lakes. The depth and breadth of knowledge and experiences that can be gained and shared is incredible!

What is even more exciting is that knowledge can lead to action. No one needs to do everything but if we do what we can, when we can, the collective actions make a difference and can help the planet too!

Still not sure what the book is about? Visualize an old, faded recipe card that is well-loved through generations of your family. Did a child stand on a chair and help mix? Did you take it to a potluck and someone asked for the recipe? Did you ever watch a cooking program and were surprised to learn that another region of the country makes a variation of your recipe? **Small Hands Can Help** gives you the basic "ingredients" to get started and helps you to nourish your curiosity and compassion for your local ecosystem, and ultimately the planet. Have fun creating your own personal ecological recipe to enjoying the outdoors and helping planet Earth.

How to Navigate This Book

This book is designed to guide families, educators and communities in happy ways to help heal our home, one small step at a time. It is not a book that you read once, cover to cover and then never pick it up again. It is designed to be read in parts, depending on the age level of the readers. As families and readers grow and mature, the book can be referenced again and again for a deeper dive into facts, programs and fun volunteer opportunities. It is a timeless representation of non-fiction paired with adventure and an excitement about all things natural.

Indicators - This book showcases eight *indicator* species, that demonstrate how ecosystems are changing. Indicators are species that signal changes in the environment. New weather patterns are accelerating the Earth's current warming trends. The indicators chosen for this book are local to the state of Minnesota. Some are beloved, while others may seem obscure. The truth is that all of them are incredibly important to this state's changing ecosystems and we will all need to work together to help them continue to thrive in this state.

Species Range Maps - On the first page of each chapter you will find a map that shows the range of the animal in that chapter. Each one shows where they can currently be found, where it used to be found (in addition to where it is currently found) and where it has never been found. The only map that is different from this pattern is chapter 6. This map is different because this species has population control practices already in place.

QR Codes and More - Each chapter includes QR codes, fact-filled poetry and conservation resources that can help you learn about the species and what you can do to help conserve it. The QR codes placed with the text will need to be scanned by a reader on a device, such as a phone. These codes will lead to species sounds, citizen science apps, conservation programs and volunteer opportunities that are reliable, fun and safe for the whole family.

Citizen Science Apps - These apps invite the general public to participate with scientists in simple easy steps. This is a very cost-effective method for scientists to gather information from all over the world, in massive quantities. It far surpasses what trained scientists can do on their own, but the data it provides to them is priceless. It helps them to learn which species are doing okay, and which are in need of help.

For example: Imagine stepping out into your backyard or local park. You hear a bird chirping in a tree. You don't know what it is so you open the *Merlin Bird ID App* by the Cornell Lab of Ornithology, and record the sound it makes. The bird is identified by the app. Later, you learn that you can connect the Merlin Bird ID app to the eBird app, which is also put out by the Cornell Lab of Ornithology. Working with the two apps together, you can follow the migration patterns of millions of birds, including the one you found at home. It's easy, a lot of fun, and can be an incredible tool to learn more about the birds in your area!

Vocabulary - On the inside of the back cover of the book you will find a vocabulary crossword puzzle with a QR code of its own. Scan that code and it will allow you complete the puzzle online. The solution to the puzzle is also found through that QR code link. Anyone can also make copy from the last page. It is a fun way to learn about the words found in bold throughout the ecosystem texts.

Sparks of Thought - Also in the back of the book you will find critical thinking questions designed for different age levels. These are open-ended questions designed to make one think about what they know and what they have learned.

Small Hands Can Help! When all of the little things are put together, they equal a big difference! No one needs to do everything, but if we all do what we can, when we can, the rewards will be fun and help the planet too! It can lead to new adventures for grandparents, families, friends, and whole communities working together or just one person at a time.

Chapter 1 Minnesota

Where Do I Roam?

By Lora Wieman

I live in this State and call it my home.
See the counties in green to know where I roam.

I have many friends but let's make it clear,
That our habitat shrinks a bit more every year.

Our lakes, wetlands, rivers, native prairies and trees...
There used to be more than you could possibly see!

Now not much is left and it gets smaller each day
But with your help maybe some could be saved.

Though the problems seem big, and your hands may seem small,
Each tiny effort will be good for us all.

We need you to help us, we need a good friend.
So don't give up reading 'til you get to the end.

Maybe you've noticed my silhouette...
Do you know who I am? Have you guessed yet?

You can learn all about me.
There's so much to know.

Join me on this adventure.
Hop to the next page and let's go!

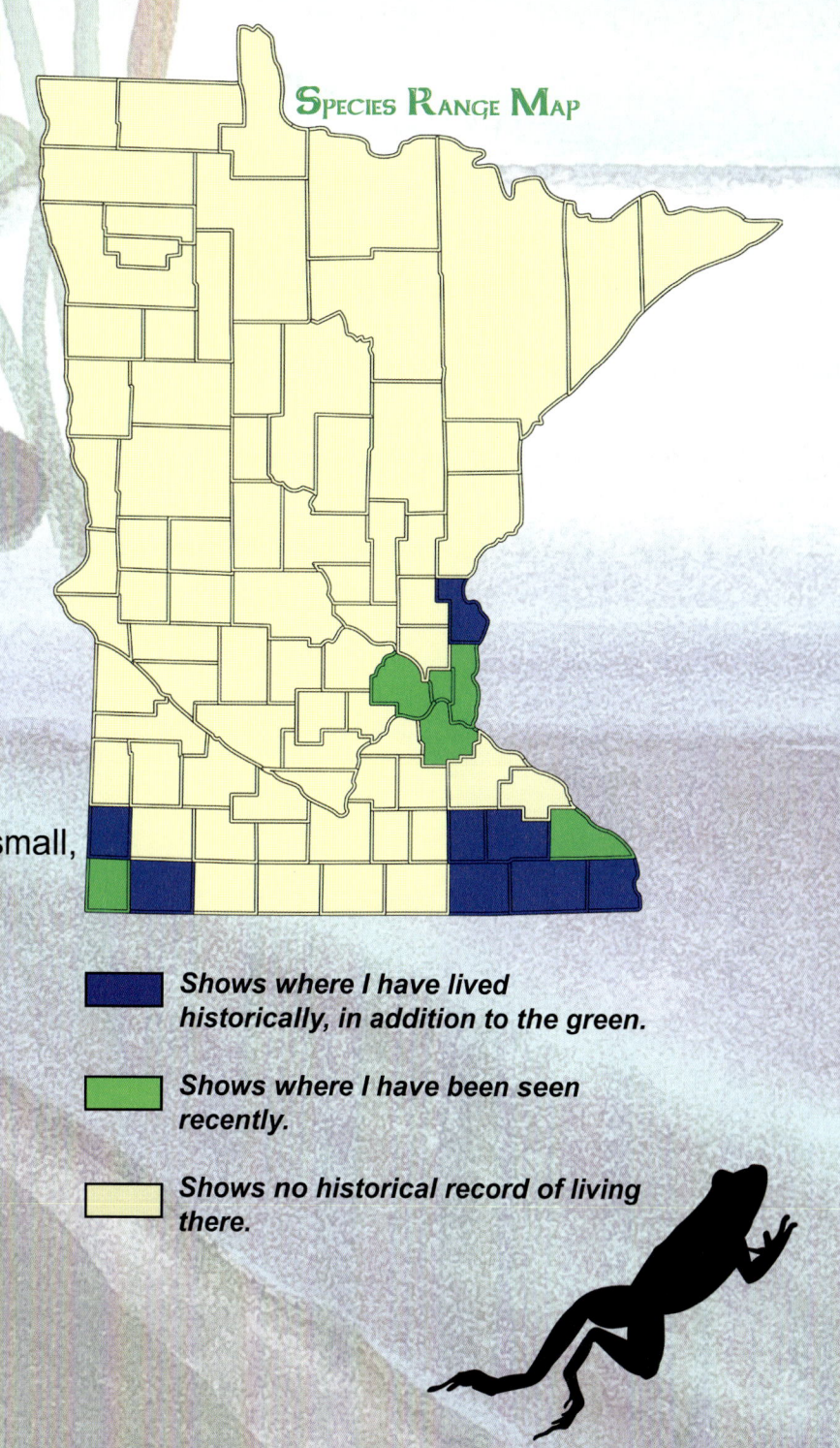

Species Range Map

Shows where I have lived historically, in addition to the green.

Shows where I have been seen recently.

Shows no historical record of living there.

BLANCHARD'S CRICKET FROG

By Lora Wieman

Scan this green QR code to listen to how different the Blanchard's Cricket frog sounds!

umesc.usgs.gov/terrestrial/amphibians/armi/frog_calls/cricket_frog.mp3

Although I'm a tree frog, I don't like to climb.
On the ground is where I spend most of my time.

I like to live along rivers and streams,
By small lakes and wetlands, the shoreline is my dream.

These are the places where you can come look,
But you may not see me outside of this book.

As an adult, I am still very tiny.
Just an inch or so long. No wonder you can't find me!

Another reason finding me is no piece of cake,
Because when you're asleep is when I'm awake.

It is then you might hear my wondrous call.
I sound like a cricket even though I'm a frog.

If you scan the code at the top of this page
You'll hear my call and you'll be amazed.

I have lots more tricks hidden up my sleeve.
The length I can jump, you will not believe!

If you could jump like me, it would be nifty,
You could start at the end zone and jump to the fifty!

Just one hop would do it. That's all it would take.
That's how I escape from birds, fish and snakes.

When I am threatened, I can jump or I hide.
But when cornered, a little chemical comes out of my side.

It sure doesn't taste good but that's why it's there.
To protect me from harm, predators BEWARE!

If they catch me I'm not a lot to eat.
I weigh the same as one raisin! Not much of a treat.

Winters are cold. Some of us don't make it through.
But I sure hope I can depend on you,

To do what you can to help keep me alive.
A frog's life is hard but with friends we'll survive.

Blanchard's Cricket Frog Ecosystem

Frogs are such a wonderful and important part of our **ecosystem** that when researchers need to check the quality of life in a certain area, one of the first things they will look at are local frog populations. The unique thing about frogs is that they live both on land and in the water. Therefore, a healthy frog population can indicate a healthy ecosystem. A frog's skin can easily absorb harmful substances that flow from adjoining lands to the water where they live. Mutant frogs or a weak **population** can indicate **toxins** or other problems that could eventually affect us all. Healthy frog populations speak loudly to the health of an ecosystem and this frog's song is music to our ears.

The Blanchard's Cricket frogs were listed by the Minnesota Department of Natural Resources (MN DNR) as an **endangered** species in 1996. They are the most **aquatic** of all North American tree frogs. The **habitat** of these frogs has been divided by our many land uses. This creates fragmented breeding conditions. This means they can only sing and hop with the other frogs from the same pond or **wetland**. They need more frogs to maintain a healthy population.

Changes in land use and other disturbances can directly affect this ecosystem, including these sensitive frogs. Water **runoff** from cropland or even our backyards can carry **sediments**, fertilizers, and chemicals applied to manage weeds or improve farm productivity. This runoff can concentrate in ponds, wetlands, and other water habitats where these frogs live. Weather events like heavy rain storms are also changing in ways that increase this effect. Changing weather patterns have other trends too. Higher year-round temperatures are causing more frequent drought conditions as well. These changes add additional stressors to this ecosystem.

Native plantings in farm waterways, or around shorelines of wetlands and ponds can buffer the effects of this harmful runoff. Purchasing organic products encourages the producers to avoid using many inputs that can cause water quality problems. If we buy locally, we can support our farm neighbors while helping protect local habitats and improve water quality.

In the end these actions will benefit both our planet and us, as well as the frogs! Helping the frogs, helps your family stay healthy too!

Little Things, Small Hands Can Do, To Make a Big Difference.

Become a volunteer frog surveyor! You can participate in the Minnesota Cricket Frog Survey! This is a community science project with collaborative work by the MN DNR Non game Wildlife Program, the Amphibian and Reptile Survey of Minnesota and you! *Black QR code - dnr.state.mn.us/eco/nongame/projects/cricket-frog-survey*

Stay on marked paths! When visiting parks, nature preserves, and important habitat areas. Don't step on a frog's house!

Learn about this frog's habitat and nature! Follow the olive green QR code to the Minnesota DNR Rare Species Guide to find out other exciting facts for the Blanchard Cricket Frog and what can be done to help preserve this species in our state. *Lime QR code - dnr.state.mn.us*

Trash Talk! Don't litter, it's deadly to animals! Never leave your garbage on the ground, streets, waterways or beaches. Our human trash can get stuck in wildlife airways or get caught in their limbs and make them unable to walk or move. Make sure you put trash where it belongs, but also make sure to recycle anything that you can. Our landfills are filling up! It's more important than ever to reuse, reduce, recycle and repurpose as much as we can.

Identify your finds with the SEEK app! Did you find a frog, bug, flower or mushroom and you are not sure what it is? Open up the Seek camera in this app to see if it can help you! Kid-safe, fun and free for families, there is no registration involved, and no user data collected. Seek will ask permission to turn on location services, but your location is obscured to respect your privacy while still allowing species suggestions from your general area. Your precise location is never stored in the app or sent to iNaturalist. This is a great app for families who want to spend more time exploring nature together.
Green QR code - inaturalist.org/pages/seek

Chapter 2 Minnesota

Where Do I Roam?

By Lora Wieman

I live in this State and call it my home.
See the counties in green to know where I roam.

I have many friends but let's make it clear,
That our habitat shrinks a bit more every year.

Our lakes, wetlands, rivers, native prairies and trees...
There used to be more than you could possibly see!

Now not much is left and it gets smaller each day
But with your help maybe some could be saved.

Though the problems seem big, and your hands may seem small,
Each tiny effort will be good for us all.

We need you to help us, we need a good friend.
So don't give up reading 'til you get to the end.

**Maybe you've noticed my silhouette...
Do you know who I am? Have you guessed yet?**

**You can learn all about me.
There's so much to know.**

**Join me on this adventure.
Flutter to the next page and let's go!**

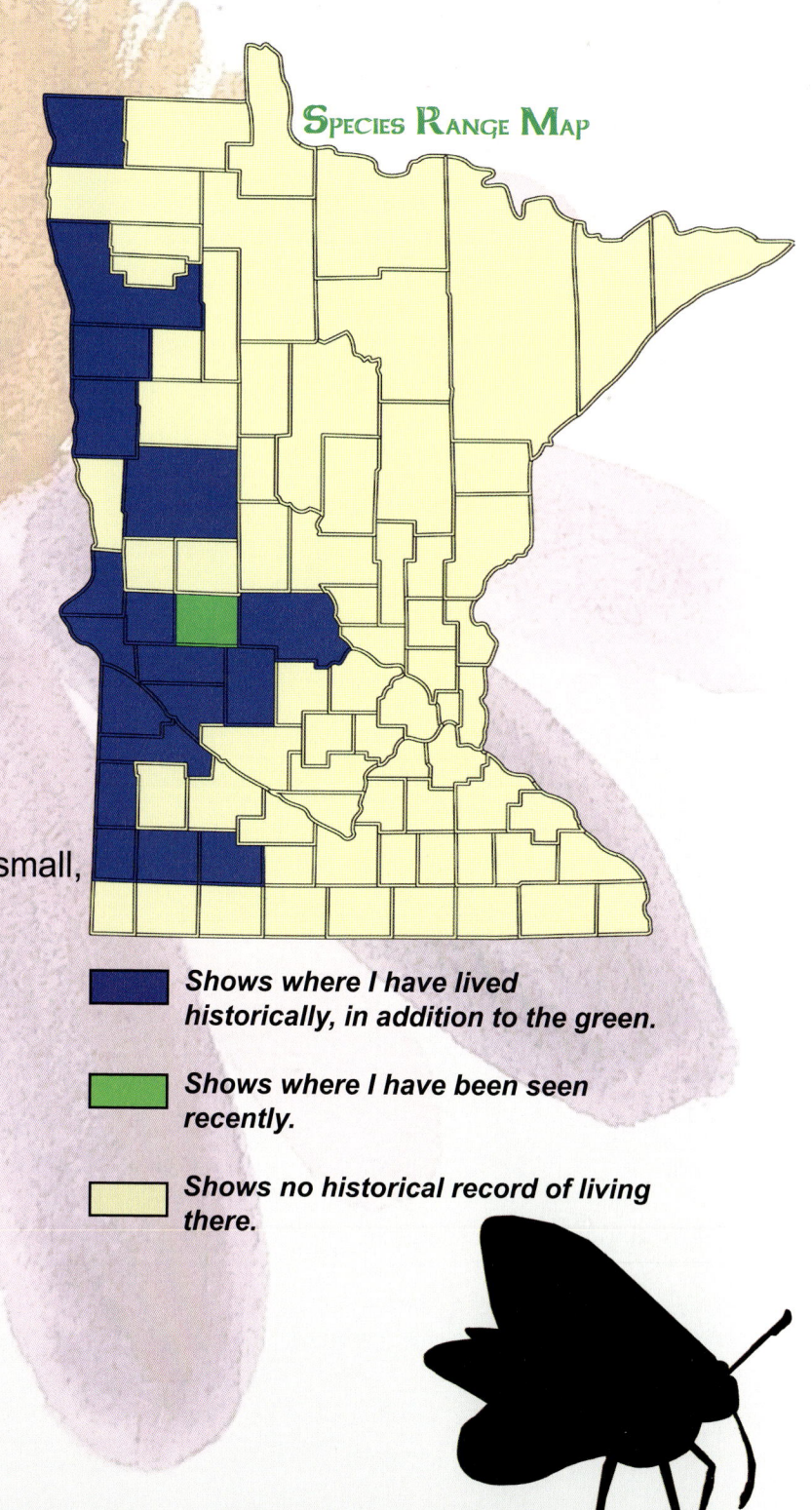

Species Range Map

■ Shows where I have lived historically, in addition to the green.

■ Shows where I have been seen recently.

■ Shows no historical record of living there.

Shine a light,
　Shine a light
　　For all to see

Our world is changing
　Right before
　　You and me

DAKOTA SKIPPER BUTTERFLY
By Lora Wieman

Have you seen my habitat?
I've looked but don't know where it's at.

All I see are shopping malls,
Gas stations and parking stalls.

If I fly outside of town,
There's machines digging up the ground.

People planting things I cannot eat
Or replacing prairie with another street.

I need a place where native grasses grow
And flowers bloom. It's what I know.

Although some disturbances are key
Grazing, small fires and haying, all help me.

Too many and my life will be at risk
It's a balance! Do you get the gist?

I'm a Dakota Skipper Butterfly.
You may not notice as I flutter by.

Or see me sitting on this twig.
I'm only an inch, my wings aren't big.

But they beat so fast as I fly
They're just a blur to the human eye.

My colors are dull: orange and brown
Not the most famous butterfly around….

But still I matter all the same
And saving me is not a game.

I have a special job to do–
I'm a pollinator! Yes, that's true!

Plants need their pollen to be shared
And some can't release it into the air.

So they wait for me to arrive,
That's how I help them to survive.

I only live for three short weeks,
In the summertime, where I will seek,

Prairie dropseed where I can lay my eggs.
Where they'll be safe. And so I beg,

For you to just please understand.
Don't take away all my land.

Scientists have figured out
That though I'm small and cannot shout,

I still may have a lot to say.
By studying me they might find a way,

To restore habitat to its best.
Cause what's good for me, is good for the rest.

All prairie dwellers; lightning bugs,
Birds and bees, badgers and skunks.

Most prairies have all gone away.
Just one percent is left today.

If you'd like to lend a hand,
Plant coneflowers on your land.

And little bluestem- a native grass so sweet.
It's what my babies like to eat.

So when you have a choice to make,
Please think of me. *Our* future is at stake.

Dakota Skipper Butterfly Ecosystem

Hilltop prairie is the most endangered ecosystem in North America. Dakota Skipper Butterflies are an *indicator* species associated with this rare and declining *community* of plants and animals. Every part of this butterfly's life depends on a different native prairie plant. These butterflies were listed as endangered by the State of Minnesota in 1984 and are federally listed as *threatened*.

Pollinators include butterflies, moths, beetles, native flies, and more than 450 species of native bees living in Minnesota! All play a key role in taking pollen from one plant to another so that new plant seeds can be created. Many food crops and native plants rely on this process to survive. But pollinators have significantly declined worldwide in recent years.

Scan the green QR code to the right, to watch a 12 minute video that shows how the Minnesota Zoo and the Minnesota Department of Natural Resources are working together to preserve and protect these butterflies and our native prairies.

The Dakota Skipper is an important pollinator for native plants. Before Minnesota became mostly farm country, Dakota Skippers were an important part of the naturally expansive prairies. The small patches of habitat that remain are commonly surrounded by cropland that largely depend on mainstream farming practices. Chemicals sprayed on these fields can drift with the wind into natural areas and kill native plants that Dakota Skippers and other animals depend on.

Other changes to their ecosystem also challenge the Dakota Skippers. Building new houses, digging new sand and gravel quarries, construction projects such as new wind turbines, and other land use changes pose the greatest risks to their *fragile* habitats. Grazing livestock can eat the very plants these butterflies need if put into native plant areas for pasture when the plants are flowering. Finally, there is trouble with current weather trends. Severe drought, more intense storms, and extended above-average hot days contribute to habitat loss.

Thoughtful planning of new land uses to avoid further habitat loss and rethinking the timing for pasturing livestock can achieve an ecological balance between our needs and the unique ecosystem these special butterflies call home. Farmers and *conservationists* can work together to create land stewardship plans to restore, manage, and protect valuable ecosystems on their land.

Little Things, Small Hands Can Do, To Make a Big Difference.

There's plenty you can do too! Raising your lawn mower blades can remove weed seed heads but keep old stems and decaying leaves as ground habitats for native bees and other pollinators. Create a patch of native wildflowers, grasses, shrubs, and trees – either at home or ask a park board about adding a pollinator planting at a park site.

Plant native wildflowers, grasses and trees! Native plants that are original to a certain region create a unique harmony with the native wildlife. It is something that develops over thousands of years. If you plant a variety of flowers native to your area that bloom in the spring, summer, and fall, it will provide nourishment and cover for the pollinators in all stages of their life.

Plant a BUTTERFLY GARDEN! Scan the dark blue QR code to the left to go to the University of Minnesota Extension web page to learn how to create a butterfly garden. *Blue QR code - iextension.umn.edu/landscape-design/creating-butterfly-garden*

Local Minnesota Butterfly Garden Resources! Lists of local Minnesota growers and vendors that will provide the right native trees, plants and seeds, for the location you need them. *Pink QR code - monarchbutterflygarden.net*

Tell your friends and family how they can help too! We need everyone to help save the pollinators!

Join iNaturalist and become a citizen scientist! Become a community scientist who helps researchers collect data about pollinators and their habitats. This app connects with the *seek* app to bring your findings to the scientific world! *Lime green QR code - inaturalist.org*

Chapter 3 Minnesota

Where Do I Roam?

By Lora Wieman

I live in this State and call it my home.
See the counties in green to know where I roam.

I have many friends but let's make it clear,
That our habitat shrinks a bit more every year.

Our lakes, wetlands, rivers, native prairies and trees...
There used to be more than you could possibly see!

Now not much is left and it gets smaller each day
But with your help maybe some could be saved.

Though the problems seem big, and your hands may seem small,
Each tiny effort will be good for us all.

We need you to help us, we need a good friend.
So don't give up reading 'til you get to the end.

Maybe you've noticed my silhouette...
Do you know who I am? Have you guessed yet?

You can learn all about me.
There's so much to know.

Join me on this adventure.
Swoop to the next page and let's go!

Species Range Map

Shows where I have lived historically, in addition to the green.

Shows where I have been seen recently.

Shows no historical record of living there.

Shine a light,
    Shine a light
        For all to see

Our world is changing
    Right before
        You and me

RED-HEADED WOODPECKER

By Lora Wieman

If you have seen me, you won't forget my head of brilliant red.
Did you know I like to nest in trees already dead?

I like to perch in mighty oaks. Acorns are what I eat.
And bugs, I catch them in midair! Some fruit is such a treat.

My wings, when I spread them out, are more than a foot wide.
But just three ounces is what I weigh. Not much! Are you surprised?

If I was heavy I couldn't fly so high and far and fast.
Though summers here are nice and warm, we know that they won't last.

So winters find me further south. I visit other states.
Like Florida and New Mexico, where through the winter I wait.

When spring comes back around again, I fly my way up north.
Watch for me in big oak trees, flitting back and forth.

If you don't see me, you might hear my strange and unique call.
"Quee-ah" is what I say to my friends one and all.

My plumes are pretty, black and white, which leads to some nicknames.
Flag Bird and Jelly Coat, are some of my claims to fame.

But most of all, I'm simply known as the Red-Headed Woodpecker
And those dead trees that I mentioned before are where I like to find shelter.

Shelter for me is getting scarce, especially near a city.
When people want to build a house, they don't think dead trees are pretty.

So all of them are chopped right down, most of the living ones too!
Now I have a home no more and have to search elsewhere for food.

The fossil records show I've been around for 2 million years.
Don't let this be the century where my existence disappears.

Scan this red QR code to listen to a Red-Headed Woodpecker!
.bird-sounds.net/red-headed-woodpecker

Red-Headed Woodpecker Ecosystem

According to the National Audubon Society: "3 billion birds have vanished in less than one human lifetime. Two-thirds of North American bird species face the risk of climate extinction. 70+ species of birds have lost half of their population in the past 50 years."

Red-headed Woodpeckers have been in decline, in Minnesota, since 1966. This once common Minnesota bird is no longer common, having dwindled by half! In 2015, Red-headed Woodpeckers were listed by the MN DNR as a "Species of Greatest Conservation Need."

Researchers have found that the lack of dead trees, used for nesting, along with fewer bugs and berries, due to agricultural chemical application, have contributed to fewer of these beautiful red-caped birds. The preferred habitat of this woodpecker is **oak-savanna** which consists of large, older and scattered oak trees with native grass and flower understories. Oak savannas are one of the most threatened plant communities in Minnesota due to human disturbance and the fatal, oak wilt disease. An oak tree is considered a **keystone species** since it can serve as shelter and homes for many animals, including being a host to butterflies. The acorns provide food for a multitude of animals. Red-headed Woodpeckers cache (store) acorns which helps in the seed-dispersal of the majestic oaks. The cavities they create for nesting are often used by other cavity-nesting birds and mammals, including chickadees and flying squirrels. Red-headed Woodpeckers are an indicator species of a healthy oak savanna. Less than 0.02% of high quality oak savanna habitat remains in the Midwest today.

The Red-headed Woodpecker is unique in a few ways. They do not peck the bark to trap insects in the sap of trees, rather they catch bugs in flight. When they cache their food they cover it with pieces of wood, insects even get jammed under bark. They will also eat other bird eggs and the occasional mouse! Providing a bird feeder with suet, seeds, corn and fruits can help bring these birds in for closer observations.

The drum of this red-headed bird has echoed through the oak savanna for generations; leading it to be known as the "spark bird" since it is often one of the first birds people can identify. We can help oak savanna habitats by keeping oak firewood local and covered to prevent spreading oak wilt, join citizen scientist organizations to raise your awareness or maybe even volunteer to help with managing oak savanna, like removing invasive species or constructing brush piles where food species of the Red-Headed woodpecker nest. May the generations to come know the swoop of black and white when it passes overhead!

Little Things, Small Hands Can Do, To Make a Big Difference.

Get Involved! The Red-headed Woodpecker Research and Recovery Project is a program of the Audubon Chapter of Minneapolis and the Cedar Creek Ecosystem Science Reserve. Since 2008, citizen scientists have monitored Cedar Creek's oak savannas to learn more about red-headed woodpecker nesting and breeding behavior.
Red QR code - rhworesearch.org

Join Audubon! Plan a visit to an Audubon Center or Sanctuary, or connect with your local Audubon chapter to explore the birds in your community.
Yellow QR code - audubon.org

Download the Merlin app and identify birds, by their sound, or a photo! Merlin is a free app put out by the Cornell Lab of Ornithology and is designed to help people find the answer to "what's that bird?" It features, "Sound ID" which listens to the birds around you and shows real-time suggestions for who's singing. Sound ID works completely off-line, so you can identify birds you hear no matter where you are. Snap a photo of a bird, or pull one in from your camera, and "Photo ID" will offer a short list of possible matches. You can also build a digital scrapbook of your bird sightings with "Save My Bird". Tap "This is my bird!" each time you identify a bird, and Merlin will add it to your growing life list. As you learn this app and grow your birding skills, you can later pair this app with the eBird app. Learn more about eBird under Chapter 5.
Black QR code - merlin.allaboutbirds.org

Chapter 4 Minnesota

Where Do I Roam?

By Lora Wieman

I live in this State and call it my home.
See the counties in green to know where I roam.

I have many friends but let's make it clear,
That our habitat shrinks a bit more every year.

Our lakes, wetlands, rivers, native prairies and trees...
There used to be more than you could possibly see!

Now not much is left and it gets smaller each day
But with your help maybe some could be saved.

Though the problems seem big, and your hands may seem small,
Each tiny effort will be good for us all.

We need you to help us, we need a good friend.
So don't give up reading 'til you get to the end.

Maybe you've noticed my silhouette...
Do you know who I am? Have you guessed yet?

You can learn all about me.
There's so much to know.

Join me on this adventure.
Swim to the next page and let's go!

Best 10 Stocked Lakes Map

The Big 10

1. Cass Lake
2. Kabetogama Lake
3. Lake of the Woods
4. Leech Lake
5. Mille Lacs
6. Lake Pepin
7. Rainy Lake
8. Upper Red Lake
9. Vermilion
10. Winnibigoshish

Shine a light,
Shine a light
For all to see

Our world is changing
Right before
You and me

WALLEYE

By Teresa Veraguth

In big, blustery Minnesota lakes, with gravelly sandbars,
There lurks the wiley Walleye, who is a fishing rockstar!

It's so popular with anglers, its called Minnesota Gold.
To catch it you must be willing to brave the dark or the cold.

When temps go low, the lakes freeze over turning snowy white,
Folks round here head to the darkhouse for a MN Date Night.

It doesn't happen everywhere, it's a unique way to court,
But here ice fishing is a *very* popular sport.

Winter is fun, but you can fish them year-round,
You just need to know where the walleye are found.

Its such a big deal here there are *two* towns, Garrison & Baudette,
Who claim to be the Walleye Capital of the World, You BET!

If you've ever eaten a Minnesota fish fry,
Then you've likely had a try of the scrumptious walleye!

Walleyes are beautiful, olive and gold, with brassy flashes and
Long, thin, torpedo-like bodies with sporty black dashes.

The walleye is a member of the freshwater perch clan.
With two jaunty fins on its back, spread in rays, like fans.

Named for its pearly eyes, made by a reflective layer of pigment,
The *tapetum lucidum* gives it the necessary equipment,

To be able to see where it isn't very bright,
So the best time to catch them is usually at twilight.

When the sun is out, they swim way down low,
To the deeps and bottoms, where the light cannot go.

But in the evening dusk and through till dawn,
They climb up from the depths until night is gone.

Up to where underwater plants, rocks and natural debris,
Even dead trees, can create a dense habitat sea.

For schools of shiner minnows, zoo plankton, and worms
Leeches, insects, and snails, all water-dwellers that squirm.

These are tasty for walleye, who are strictly carnivores!
So if you plan to spend some time out-of-doors,

Drop a fluorocarbon line! Lures are fun, but cannot compete,
With leeches, worms and minnows - the walleye's sweet treat!

Now that you know the right bait and the "when",
Look to the chart to see the "The Big 10"!

In many Minnesota lakes, walleye are both natural and stocked,
But The Big 10 is where anglers have flocked.

It doesn't matter if you are a tiny tot or an angling master,
Nothing gets your blood pumping faster,

Than seeing the shine, like the minors of old,
While ice fishing at your special hole & spying a flash of GOLD!

A true fighter with sharp teeth, it is line shy, and tricky to catch,
It's the ultimate prize for an angler able to outmatch!

If you would like to take the time to try and goldmine.
Minnesota law requires a fishing license for anyone 16-89!

Walleye Ecosystem

The Minnesota DNR estimates that across the state anglers generate several billion dollars of revenue. Ecotourism including hotels, restaurants, travel, and gear provides more than 35,000 jobs. Funds raised from fishing license sales help to support many environmental causes in the state, including the crucial management of shorelines.

The walleye is the most *iconic* fish in Minnesota! When it is found in rivers, it is native to those waters. The Minnesota DNR has fisheries located in various parts of the state. Walleye are stocked in specific lakes. This increases the availability of this much sought after fish.

About 85% of walleye, in Minnesota, naturally reproduce but there is some evidence this number may be declining. Warming climates have shortened ice cover which impacts spawning and water temperatures have risen above the 65-70° F, which is ideal for these cold-water species. Algae blooms lower oxygen levels through the process of eutrophication and invasive rusty crayfish eat their eggs.

Preventing the spread of *invasive* species is something all anglers and watercraft enthusiasts can do to protect our celebrated waterways:

- Carefully inspect and remove water plants and invasive animals, like zebra mussels, from all forms of watercraft and the trailers used to haul them.
- Drain water from all areas including live bait wells. Leave the drain plugs out after draining. It is illegal to release aquatic animals or plants from one body of water into another.
- Do NOT dump unused live leeches or worms into the waters or woods. Non-native worms have been known to create an entirely different soil ecosystem, this reduces the ability of the soil to soak in rain and snow-melt, leading to more erosion of soil into waterways.
- Replace all tackle containing lead with nontoxic alternatives such as tin, bismuth, steel, and tungsten-nickel alloy. Loons and other waterfowl, like swans, die from consuming lead tackle found on the bottom of lakes and eating fish with lead tackle still in them. The enjoyment of watching these birds glide past your boat is a pleasure that everyone should have the opportunity to experience.
- ALWAYS know the rules and regulations that apply to the species of fish and the area you are fishing.

"All walleye that are 18 to 26 inches in length, inclusive, must be immediately returned to the water. A person's possession and daily limit for walleye is four, and must not include more than one walleye over 26 inches in length." - MN DNR. Scan the olive green QR code to learn best practices for catch and release.

Little Things, Small Hands Can Do, To Make a Big Difference.

Wild Spotter™ - Engaging and empowering the public to help find, map, and prevent invasive species in America's wilderness areas, wild rivers, and other natural areas. Become a Wild Spotter citizen scientist volunteer, download the Mobile App, and help protect America's Wild Places! *Blue QR code - wildspotter.org*

Rajala Woods Foundation. Support and volunteer for organizations that care about recreating Minnesota's native ecosystems. This is a non-profit organization devoted to restoring native forests, conserving natural ecosystems and ensuring public access for recreational use and enjoyment. *Yellow QR code - rajalawoodsfoundation.org*

EDDMapS. This app brings data together from other databases and organizations, as well as volunteer observations to create a national network of invasive species and pest distribution data that is shared with educators, land managers, conservation biologists, and beyond. This data will become the foundation for a better understanding of invasive species and pest distribution around the world. *Turquoise QR code - eddmaps.org*

FISH ID! Down load the app on your phone and take it with you! This fish identification tool was developed by the University of Wisconsin Center for Limnology, Wisconsin Department of Natural Resources, and the University of Wisconsin Sea Grant Institute. (Recommended by the MN DNR) The app contains thousands of fish photos so it is nearly 300Mb. Be sure you are connected to a wi-fi network before downloading. *Salmon colored QR code - seagrant.wisc.edu/fish-id/*

Chapter 5 Minnesota

Where Do I Roam?

By Lora Wieman

I live in this State and call it my home.
See the counties in green to know where I roam.

I have many friends but let's make it clear,
That our habitat shrinks a bit more every year.

Our lakes, wetlands, rivers, native prairies and trees...
There used to be more than you could possibly see!

Now not much is left and it gets smaller each day
But with your help maybe some could be saved.

Though the problems seem big, and your hands may seem small,
Each tiny effort will be good for us all.

We need you to help us, we need a good friend.
So don't give up reading 'til you get to the end.

Maybe you've noticed my silhouette...
Do you know who I am? Have you guessed yet?

You can learn all about me.
There's so much to know.

Join me on this adventure.
March to the next page and let's go!

Species Range Map

Shows where I have lived historically, in addition to the green.

Shows where I have been seen recently.

Shows no historical record of living there.

Shine a light,
Shine a light
For all to see

Our world is changing
Right before
You and me

PIPING PLOVER

By Angela Coyer & Teresa Veraguth

If you are in northern Minnesota, near the waters and shores,
Take some time, slow down, and rest your oars.

You might just see a wing of special birds fly over!
It's an endangered species, known as the Piping Plover.

In springtime our shores become their nesting and breeding grounds
It's an amazing thing to view, the males marching like soldiers and making funny sounds.

Males will search the beaches to find a spot that is the best,
Then begin several nesting scrapes to attract a female who will finish the rest.

The nests are shallow, built on the beach, the ladies will finish one, with shells and pebbles they find.
Both mom and dad take turns, warming the eggs that she lays inside.

The eggs look like the rocks in the sand,
Blending in with, and camouflaged, by the land.

Eggs are lain with no more than four to a batch,
Twenty-eight days later, the chicks can walk, run and feed, as soon as they hatch!

The female will leave the nest before the young fledge or are able to fly,
The male will remain to raise the hatchlings, alone, peeping a piper lullaby.

If Dad feels the chicks are threatened, he will pretend he has a broken wing,
Then lead the intruder away from the nest to protect his tiny offspring.

During the Minnesota summers, they are found enjoying the beach and it's vegetation.
But in late July, they gather in flocks, to form a "brace" for migration.

Together they will take the long journey south, to the Gulf and Atlantic Coast.
They can't survive Minnesota in the winter so they fly south, to find the food they need the most.

While the average lifespan of a piping plover is 5-6 years, on a Great Lakes beach,
There are two males and one female banded and recorded at 15 years each.

Piping plovers are beautiful creatures,
Maybe you'll see one fly your way

Let's do something together
To save and protect this endangered species today!

Piping Plover Ecosystem

The Piping Plover was designated an endangered species, in Minnesota, as early as 1984. The U.S. Fish and Wildlife Service listed them as federally or threatened, throughout their range, in 1986. Unfortunately, this small shore bird's population has continued to struggle.

The Piping Plover is an indicator species because it allows researchers a window into the health of the ecosystem. Piping Plover nest some distance from the shoreline where large rocks, shoreline grasses and sedges are found. This vulnerable area is also the home to other species of birds, invertebrates and rare plant species. The Piping Plover even feeds on some of the shore flies and midges that are irritating to beach goers. This benefits **ecotourism,** bringing people to view this secretive little plover, or to enjoy less pests on a cool, windy beach.

The Piping Plover is currently found in one region in northern Minnesota, Lake of the Woods County! Morris Point, Garden Island, Pine & Curry Islands and the Rocky Point Wildlife Management Area are locations where lucky observers may still spot one bursting across the sand after a beetle. The Piping Plover has become so rare for several reasons: climate change, shoreline development, off-road vehicles and predation. Intensive efforts by government and local conservation groups have been made to protect nesting habitat, often by placing plastic enclosure fencing around nest sites. Where management of shoreline habitat has been conducted, there are often signs to explain the importance of these areas for habitat, to reduce erosion, filter runoff, and increase climate resilience from ice damage and severe rainfall events.

It is up to all of us to do our part! **Erosion** increases when shoreline development does not protect the sensitive grassy edges or when we trample through these areas, dragging lawn chairs or coolers behind. A severe infraction occurs when off-road vehicles tear right through the shoreline. Dogs joining us at the beach should be kept out of the shoreline grasses and on a leash.

ALWAYS stay on the designated paths and trails when heading out for a day at the beach, including while kayaking, canoing or paddle-boarding. When camping on the shoreline, stay in the designated areas. Pack out everything, do not leave any trash behind on the beach. Trash encourages raccoons, skunks, crows, gulls, feral cats and foxes to scour the beaches for food. They often find Piping Plover eggs and chicks to be an easy treat.

Learn more about ways citizen scientists can help. You can volunteer on shoreline management days. Ask others to join you! If you are one of the few, to spot this delightful little bird, it will bring a smile to your face.

Little Things, Small Hands Can Do, To Make a Big Difference.

Participate in NestWatch! Nest watching is easy and just about anyone can do it, although children should always be accompanied by an adult when observing bird nests. Simply follow the directions on the website to become a certified NestWatcher. You can find a bird nest using the helpful tips, visit the nest every 3-4 days and record what you see, then report this information to *NestWatch*. You can also download the *NestWatch* mobile app for iOS or Android and record what you see at the nest in real time. *Turquoise QR code - Cornell Lab of Ornithology.*

Join eBird and contribute to citizen science! eBird began with a simple idea—that every birdwatcher has unique knowledge and experience. The eBird goal is to gather birding information and then freely share it, to power new data-driven approaches to science, conservation and education. At the same time, they have developed tools that make birding more rewarding. From being able to manage lists, photos and audio recordings, to seeing real-time maps of species distribution, to alerts that let you know when species have been seen, Ebird strives to provide the most current and useful information to the birding community. Minnesota eBird is a collaborative project managed by the Minnesota Ornithologists' Union. *Green QR code - Cornell Lab of Ornithology*

Check out the Questagame App! Play the world's first mobile game that takes you outdoors to discover, map and ultimately help protect life on our planet. This game is free. Your sightings contribute to real research and conservation. Follow the QR code to watch a video about how it works. *Olive QR code - Questagame.com*

Chapter 6 Minnesota

Where Do I Roam?

By Lora Wieman

I live in this State and call it my home.
See the counties in green to know where I roam.

I have many friends but let's make it clear,
That our habitat shrinks a bit more every year.

Our lakes, wetlands, rivers, native prairies and trees...
There used to be more than you could possibly see!

Now not much is left and it gets smaller each day
But with your help maybe some could be saved.

Though the problems seem big, and your hands may seem small,
Each tiny effort will be good for us all.

We need you to help us, we need a good friend.
So don't give up reading 'til you get to the end.

**Maybe you've noticed my silhouette...
Do you know who I am? Have you guessed yet?**

You can learn all about me.
There's so much to know.

Join me on this adventure.
Zip to the next page and let's go!

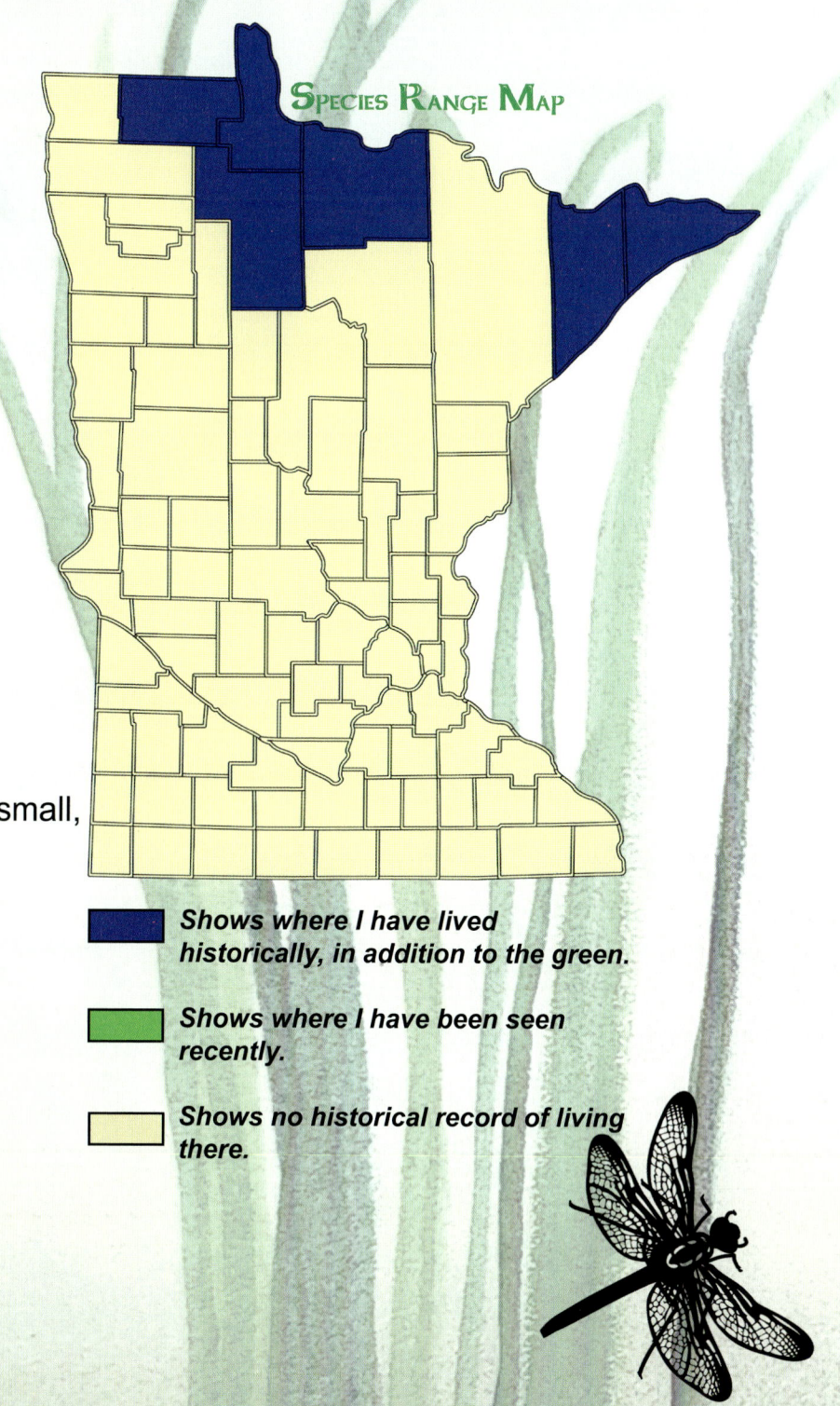

Species Range Map

■ Shows where I have lived historically, in addition to the green.

■ Shows where I have been seen recently.

■ Shows no historical record of living there.

ZIGZAG DARNER DRAGONFLY

By Rebecca Polum

Strong winds can't stop me,
I'm built for speed and strength.

I was hatched underwater,
But now I fly at length.

In all directions I can see...
Except directly behind me.

I hunt insects that bite,
With my antenna and might.

When hearts become heavy,
The dragonfly is near.

Clouds that were cluttered,
Now become clear.

His life is short,
His transformation dear.

Thank you, dragonfly,
For bringing us cheer.

Zigzag Darner Dragonfly Ecosystem

The Zigzag Darner Dragonfly has very specific needs and relies on cool water habitats in **bogs** and **fens**. It was listed as a species of special concern in 2013, by the Minnesota Department of Natural Resources. Northern Minnesota has a few remaining pockets where these magnificent darners are found. Red Lake Peatland, Mulligan Lake Peatland and Sand Lake Peatland are designated Scientific and Natural Areas to help with the conservation of this species.

Dragonflies are very important aquatic, indicator species. Scientists, and citizen scientists alike, monitor the population and locations of the **nymph** (larvae.) Depending upon the species, water temperature, and food availability, dragonflies spend as much as the first two years of their lives in the water. The nymph prey upon mosquito larvae, aquatic fly larvae (like black flies), scuds (freshwater shrimp) and other aquatic insects.

Dragonfly nymphs are sensitive to pollution. Other aquatic larvae like the caddisfly, stoneflies and mayflies are sensitive too. Surveys of the non-vertebrate species in water, are conducted to inform scientists and community leaders about the water quality in an area.

The bogs and fens of the Zigzag Darner's habitat are warming with climate changes. This could lead to the disappearance of this species in the state of Minnesota.

In addition, the MN DNR Rare Species Guide states, *"Although the peatland breeding habitats needed by the Zigzag Darner tend to be currently inaccessible and remote, there has been some recent activity hinting toward a renewed interest in peat mining in Minnesota. Such activity could severely threaten the survival of this species in mining areas. The breeding habitat is susceptible to degradation by any land use activity that changes the* **hydrology** *of the habitat or increases erosion or runoff."*

The Zigzag Darner is the smallest of Minnesota's nine species of mosaic darners, measuring 2.4 inches in length. They are a jewel to behold! The adults generally fly low to the ground or water, from early June until late September, in pursuit of adult mosquitoes, mayflies, flying ants, biting flies, and even the occasional small butterfly. They are rather unique as they frequently perch on the ground. The males have a beautiful, rich blue color with the tell-tale zigzag pattern in black; the females are a lighter blue.

Since both the nymph and adult stages of the Zigzag Darner dragonfly are carnivorous, eating mosquitoes and nuisance flies, they are nature's pest control. An adult darner can consume up to 100 mosquitoes a day! More surveys need to be done to locate other possible remote populations. Let's protect these amazing, flying pieces of art!

Little Things, Small Hands Can Do, To Make a Big Difference.

Join the Minnesota Dragonfly Society! Their mission is: *Ensuring the Conservation of Minnesota's Dragonflies and Damselflies through Research and Education.* This non-profit is looking for the Zigzag Darner (along with many other species of Odonata) throughout Minnesota. There are still many parts of Minnesota that need to be surveyed and counties that have only a few records. There are new records to find, possibly even a few new state species, and there are many people in Minnesota that still do not know how important dragonflies are to the environment that we share. *Turquoise QR code - mndragonfly.org*

Check out the amazing projects at Zooniverse! Zooniverse is the world's largest and most popular platform for people-powered research. This research is made possible by volunteers — more than a million people around the world who come together to assist professional researchers. Our goal is to enable research that would not be possible, or practical, otherwise. Zooniverse research results in new discoveries, datasets useful to the wider research community, and many publications.

You don't need a specialized background, training, or expertise to participate in Zooniverse projects. They make it easy for anyone to contribute to real academic research, on their own computer, at their own convenience. *Green QR code - zooniverse.org*

Join the Dragonfly Swarm Project via SciStarter! Everyone can participate in a large-scale study of dragonfly swarming behavior. If you see dragonfly swarms record and share your findings with the app. *Green QR code - scistarter.org/dragonfly-swarm-project*

Find this and thousands of citizen science projects on SciStarter.org!

Chapter 7 Minnesota

Where Do I Roam?

By Lora Wieman

I live in this State and call it my home.
See the counties in green to know where I roam.

I have many friends but let's make it clear,
That our habitat shrinks a bit more every year.

Our lakes, wetlands, rivers, native prairies and trees...
There used to be more than you could possibly see!

Now not much is left and it gets smaller each day
But with your help maybe some could be saved.

Though the problems seem big, and your hands may seem small,
Each tiny effort will be good for us all.

We need you to help us, we need a good friend.
So don't give up reading 'til you get to the end.

Maybe you've noticed my silhouette...
Do you know who I am? Have you guessed yet?

You can learn all about me.
There's so much to know.

Join me on this adventure.
Dive to the next page and let's go!

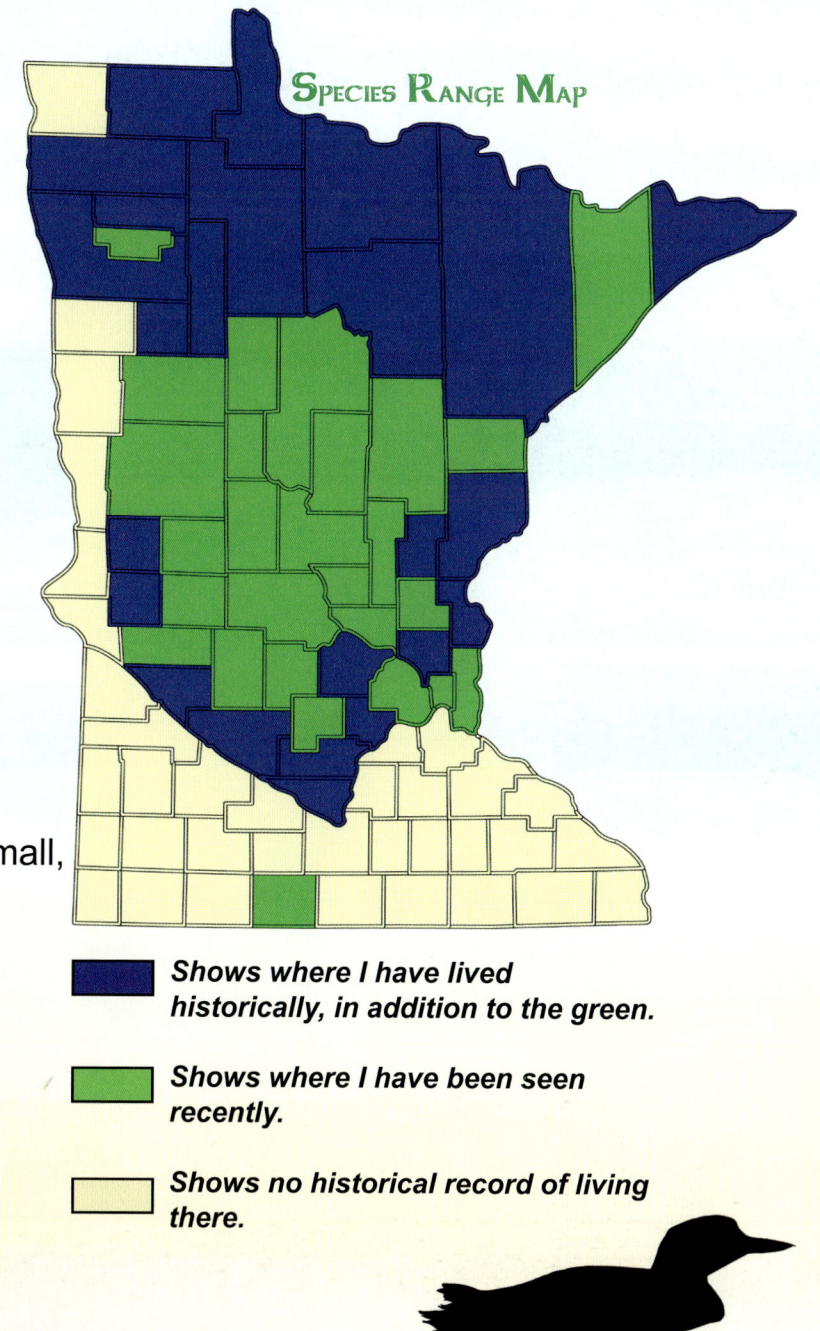

Species Range Map

■ Shows where I have lived historically, in addition to the green.

■ Shows where I have been seen recently.

□ Shows no historical record of living there.

COMMON LOON

By Teresa Veraguth

Have you ever heard a long haunting wail, from a calm lake on a warm Minnesota night?
This is the eerie and mystical sound of a loon male, searching for his mate to reunite.

In the Northwoods, loons can live for more than 30 years,
Their soul-searching calls can move a person to tears.

Loons talk with a language all their own
A wail, a yodel, a loony tremolo, a hoot or a coo, all in a different tone.

The wail is used for finding a mate or one of their young.
When males are protecting their territory a yodel rolls off their tongue

The tremolo sounds like a crazy person laughing, and can be a signal for distress
A loon will hoot or coo when they are talking to their loonlets, happy or curious.

Most birds have hollow bones, which keep them light so they can fly.
But loons have solid bones that sink like a brick and are made to dive.

Vrooooom, like a motor on a boat, their webbed feet are found in back.
Zoooooom to the bottom of the pond where they catch minnows, bugs and crawdad snacks

They build their nests close to the waters edge, so they can dive right in.
Because their feet are so far back, watching them walk on land will make you grin!

Walking on land makes them wobble and waddle, it seems kind of goony,
Its impractical and really flappable, you might even say, it looks pretty loonie!

When they dive they flatten their feathers to remove all the air, this allows them to go as deep as 250 feet.
While underwater they can both hold their breath and eat for five minutes, which is impossible for you and me!

Short wings and long heavy bodies, make take-offs tough to get them in the air
For these checkered, big birds, it takes a headwind and a long lake runway, like an imaginary set of stairs

Once they are up, they can't soar or glide, they must flap, flap, flap their wings, way more than most!
Just like a lot of grandparents from Minnesota, they head south in winter, to the salt water of the Atlantic Coast.

From Minnesota they fly south to North Carolina, to Florida, or along the Gulf of Mexico.
They can fly 75 mph and travel 3000 miles, how they remember where to go, no one seems to know!

The Common Loons of Minnesota change a lot when winter takes them away!
Their green head, magnificent red eyes and geometric checkers all turn to a drab gray.

But their red eyes return in spring as they find their way home to the same fresh water lake year after year.
At night, they find their mates with the haunting wail duets... breeding season is here.

Common loons need clean, clear lakes, dark skies and habitat without ruckus!
Together, we must protect them, or lose the echoes of the beautiful loons among us.

Yodel

Wail

Tremolo

Hoot /coo

Loon calls provided by journeynorth.org

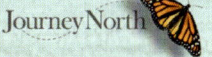

Common Loon Ecosystem

The loon ecosystem is at the greatest risk from human disturbances, which include: shoreline development, becoming tangled in fishing line, artificial light disrupting migration, near-shore watercraft washing out nests and causing stress, and water level fluctuations, due to dams, municipal runoff or climate change.

We all want to enjoy the calls and behaviors of loons for generations to come! Some lakes post loon nesting sites, people should avoid disturbing these areas. The MN DNR has found some promising results, adding nesting rafts in areas with fluctuating water levels. When fishing, be sure to retrieve and properly dispose of fishing line.

The Minnesota Pollution Control Agency (MPCA) did a study and discovered that lead poisoning was responsible for 25% of loons found dead. Older fishing tackle contains lead. It is time to change up our gear and properly dispose of the old, while replacing it with safer versions. These nontoxic alternatives are inexpensive and ecologically sound: tin, bismuth, steel, and tungsten-nickel alloy.

  *Watch this video to learn about lead and loons from the National Loon Center, in Crosslake, MN.*

Another toxic substance in the loon ecosystem is mercury. This heavy metal can **bioaccumulate** through the food chain. Loons are **carnivores** that build up mercury in their bodies and even pass this heavy metal to the eggs, reducing hatching rates. Even though mercury does naturally occur in rocks, we should always dispose of old fluorescent lights and mercury thermometers at a hazardous waste facility, not the landfill.

Loons are an iconic part of Minnesota. In 1961 they officially became the state bird. In 2005, the Minnesota State Quarter was released with a loon on the back. Then again, on June 14, 2018, The Voyageurs National Park Quarter, the 43rd coin in the America the Beautiful Quarters® Program was announced. The 43rd quarter was released on the 43rd anniversary of the founding of Voyageurs National Park and the back features a loon against the rocky shoreline of the Northwoods. Do you have a Minnesota "Loonie" in your collection?

Little Things, Small Hands Can Do, To Make a Big Difference.

 ***Adopt a Loon!*** By symbolically adopting a loon, you will help protect the Common Loon and the lakes they live in through habitat protection, education, and research.
Black QR code - nationallooncenter.org

 ***Help Birds Migrate with the Dark Sky Initiative!*** Light pollution disrupts wildlife, impacts human health, wastes money, energy, contributes to climate change, and blocks our view of the universe.
Purple QR code - darksky.org

 ***Join Minnesota's, Starry Skies North!*** People all across the region have a vision of a world free from light pollution and full of starry skies. Because of our location, we have a rare opportunity to become nationally recognized for the stewardship of the northern skies.
Blue QR code - starryskiesnorth.org

 ***Volunteer for the MN Loon Monitoring Program!*** This long-term project of the Minnesota Department of Natural Resources' Nongame Wildlife Program. Volunteers count the number of loons seen during a 10-day period in summer, and report these observations for data management & analysis.
Turquoise QR code - MN DNR

Explore Journey North! Journey North has a vision for a future where individuals of all ages create a community of action, by contributing observational data and acting as ambassadors for the conservation and protection of migratory species. *Gray QR code - Journeynorth.org*

Chapter 8 Minnesota

Where Do I Roam?

By Lora Wieman

I live in this State and call it my home.
See the counties in green to know where I roam.

I have many friends but let's make it clear,
That our habitat shrinks a bit more every year.

Our lakes, wetlands, rivers, native prairies and trees...
There used to be more than you could possibly see!

Now not much is left and it gets smaller each day
But with your help maybe some could be saved.

Though the problems seem big, and your hands may seem small,
Each tiny effort will be good for us all.

We need you to help us, we need a good friend.
So don't give up reading 'til you get to the end.

**Maybe you've noticed my silhouette...
Do you know who I am? Have you guessed yet?**

**You can learn all about me.
There's so much to know.**

**Join me on this adventure.
Amble to the next page and let's go!**

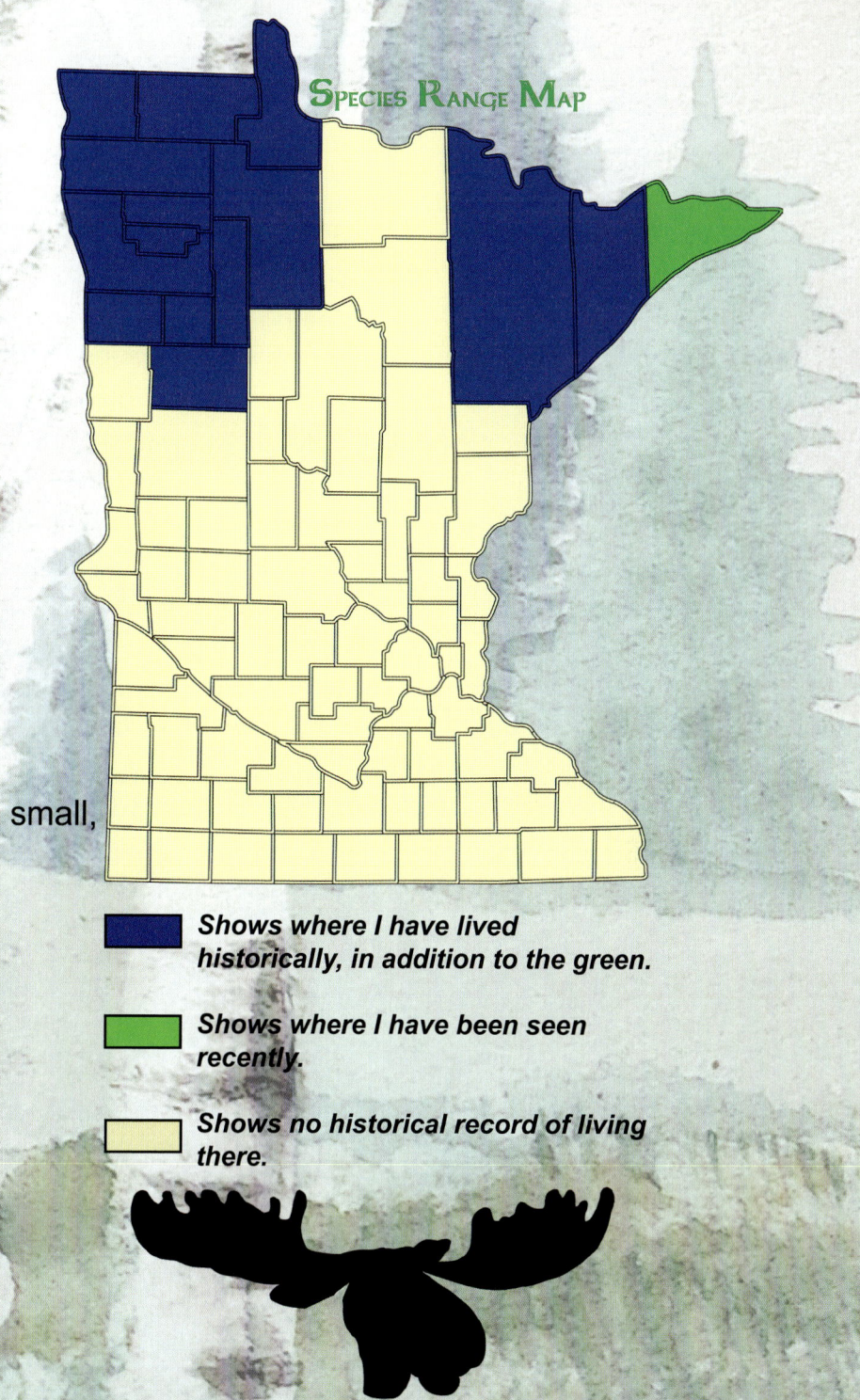

Species Range Map

■ Shows where I have lived historically, in addition to the green.

■ Shows where I have been seen recently.

■ Shows no historical record of living there.

WESTERN MOOSE
~ By Teresa Veraguth

Have you spent any time in the swamp
This is where moose like to chomp,
In the Minnesota Northwoods, near the tall, tall pine.
Streams, ponds and swamps are their most favorite places to dine

Nam, nam, nam, grass, bark, pine-cones, twigs and leaves,
Pine needles, fruit, shrubs and water plants like duckweed.
All delicious to this plant only, herbivore, who doesn't eat burgers and cheese!
They will gobble up 20 different types of plants a day, standing in water above their knees.

Their tum-tums have four different compartments, like that of a bovine cow,
This helps them digest all of the crunchy roughage of their plant chow.
Moose have special lips that can peel tree bark for their salad,
They eat for eight hours a day, 40-70 pounds of food is seriously valid!

Guess what! Just like the loon, a moose can dive, down 20 feet,
It may hold its breath for a full minute under water and EAT!
They can swim six miles per hour for up to two hours at a time,
For a land mammal with hooves those swimming skills are prime!

The huge Western Moose is anything but small
They can be up to 9 feet long and 7 feet tall.
Bulls can weigh-in at as much as 1600 pounds,
Cows are slightly smaller, but in person, their size still astounds!

Up on top are two big fuzzy ears
Out back a short stubby tail appears
Four long, skinny legs reside below
And Up front, a muzzle longer than horse's nose!

Despite their size, moose seem harmless and quiet.
Since they only eat plants, other animals aren't part of their diet,
However, they are fierce about protecting their young, so be very aware,
Deadly kicks from the hooves of a moose can kill even a grizzly bear!

Males can have humongous antlers that sprawl
They use them to push, like a wall,
Against other bull moose, in a fight to be king!
Their antlers can be 60 pounds and six feet wide, like an eagle's wings!

They fight to win the ladies' favor
The first bull to push the other back, proves he is the braver.
Winning bull becomes king of the territory.
The weaker bull moves on, the winner getting all the glory.

When mating season ends and winter sets in,
The antlers fall off, but next spring, will grow again.
The cows will give birth to one or two calves come spring,
The babes will stay with mom until next mating season is in its upswing.

Under their chin swings a bell of skin, called a dewlap, officially.
To those who study them, it is a unique feature and it's purpose seems to be a mystery.
But when they walk, the bigger it is, the bigger they swagger,
Swinging the dewlap bell back and forth like a big ol' size bragger!

Moose are the world's largest deer family,
Front legs longer than the rear, make them giant, scruffy and gangly.
A hump forms from the large muscles where their shoulders meet,
These muscles are needed to support their huge heads and uneven feet.

Moose ears and moose eyes, when uncovered,
Can move independently from each other.
The ears can rotate 180 degrees; the eyes are near-sighted and color-blind,
And, like a chameleon, swivel separately on both sides of the head, as inclined.

Moose need to live in places with long, snowy winters, because of their warm fur coat,
The hair of their fur is hollow, for insulation and to help them float.
In the warmer weather they hang out in the water to help keep them cool,
Much of their food is there and it keeps them safe, Moose are no fools.

A moose on the loose can out run a human at five days old,
Now wouldn't that be a sight to behold!
As an adult things become very fun
They can zip along at 35 mph, which is a pretty good run!

Fun fact: Moose have a nickname that is kind of wonkey!
Have you ever heard of a Rubber Nosed Swamp Donkey?
While it's meant to poke a little fun, it just shows how much they are adored,
Because nicknames are given to something or someone we have affection toward!

Western Moose Ecosystem

Moose are an important ecological link in their ecosystem. They munch their way through 40-70 pounds of **vegetation** every day, which makes them an essential part of the food chain. Moose also provide food for predators, which includes humans. Native cultures have used all parts of the moose as a source for food, clothing, shelter, medicine, tools and utensils. It's beauty, strength and majesty have been revered for generations as spiritually significant. It is a special gift to be in the presence of this huge, graceful being.

The Minnesota moose population has been cut by more than half since 2006 due to overheating, disease, and parasites—all tied to warming temperatures. When the weather is hotter, or warmer for longer than normal, moose will lie down and try to stay cool. This interrupts their **foraging** patterns and they become malnourished.

Biologists, however, say the majority of the decline of the moose population is from the increasing infection of brainworm. This parasite has become rampant from shorter winters, a direct effect of our changing climate. While grazing, moose will accidentally consume snails and slugs that have the brain worm parasite. This could lead to infection, which can damage their brain and spinal cord. The moose become disoriented and weak, loosing the ability to take care of themselves, and most will eventually die.

It is the white-tailed deer population that carries this brainworm, but deer are unaffected by the disease. **Managed forestry** programs will support a better habitat and can help keep the brainworms in check!

Warmer winters have also caused tick populations to explode, This directly affects moose numbers. An animal that is carrying a lot of ticks, will weaken from **anemia** (lack of iron from blood loss) which takes many of the herd. Ticks also leave moose more vulnerable in the winter. In an attempt to rub the ticks off, hairless patches develop on their hides leaving them without protection from the elements of a northern Minnesota winter.

"We are all concerned about the moose, and we don't want to see them disappear. Every little thing we do to help them is a good thing." - Nancy Hansen, DNR area wildlife manager in Two Harbors, MN.
dnr.state.mn.us/mcvmagazine/issues/2015/sep-oct/moose-in-minnesota
Reprinted with Permission
Minnesota Conservation Volunteer magazine, Minnesota Department of Natural Resources

Little Things, Small Hands Can Do, To Make a Big Difference.

Adopt a moose today! Every adoption kit donation helps fund the planting of trees, which creates habitat for wildlife, prevents soil erosion, and increases canopy coverage for communities. All trees from these donations will be planted in the United States and are native species that help protect and restore the wild places that nurture wildlife.
Green QR code - National Wildlife Federation.

Plant a tree! The Nature Conservancy is part of a collaborative effort to improve habitat for the state's declining moose population. Tree plantings and brush removal will eventually create a patchwork of habitats needed by moose for cover and forage. *Blue QR code - nature.org*

Avoid palm oil products! Palm oil is **unsustainable** and the use of it promotes the loss of rainforest. Trees absorb and store carbon dioxide. If forests are cleared, or even disturbed, they release carbon dioxide and other greenhouse gases. Forest loss and damage is the cause of around 10% of climate change. *Red QR code - rainforestrescue.org*

"To ensure the survival of cherished wildlife species like the moose, policies and practices are needed to address climate change. This includes reducing carbon pollution as well as adopting climate-smart approaches to wildlife conservation. We must make a serious effort to reduce carbon pollution at every level—from the choices we make in our households to the policies we adopt as a nation. America needs to embrace the development of responsible clean energy, such as wind and solar. And we must prepare for and manage the impacts of climate change to conserve our wildlife resources." -*National Wildlife Federation Wildlife Guide for Moose*

Lower your carbon footprint! Carbon dioxide emissions from driving your car, leaving your lights on, and other daily activities are making the planet hotter. Learn about greenhouse gas emissions in your city, county, state, and across the whole U.S. with Crosswalk Labs emissions map! Explore how emissions change over time and see how they differ from place to place! Visit your city's website to learn what you can do locally to lower your carbon footprint. *Dark blue QR code - crosswalk.io*

January

5 - Bird Day
10 - Cut Your Energy Costs Day
26 - Int'l Environmental Education Day
28 - Int'l Reducing CO_2 Emissions Day

February

2 - World Wetlands Day
24 - National Skip the Straw Day
Last Monday: National Invasive Species Awareness Week
nisaw.org

March

3 - World Wildlife Day
Second Friday - Solar Appreciation Day
4 - World Engineering Day for Sustainable Development
6 - World Conservation Strategy Day
12 - National Plant a Flower Day
14 - Int'l Day of Action for Rivers
15 - World Consumer Rights Day
18 - Global Recycling Day
20 - World Frog Day
 - World Rewilding Day
21 - Int'l Day of Forests
 - World Planting Day
 - World Wood Day
22 - World Water Day
Last Saturday - Earth Hour
30 - International Day of Zero Waste

April *Earth Month*

3 - World Aquatic Animal Day
17 - Bat Appreciation Day
22 - Earth Day
Last Friday - National Arbor Day

May *American Wetlands Month*

4 - Greenery Day *(In Japan)*
13 - National Windmill Day
16 - National Love a Tree Day
17 - World Recycling Day
19 - Plant Something Day
Second Saturday - World Migratory Bird Day
 - National River Clean-up Day
Third Friday - Endangered Species Day
20 - Bike-to-Work Day
 - World Bee Day
22 - World Biodiversity Day
23 - World Turtle Day
24 - European Day of Parks
29 - Learn About Composting Day

Environmental Green Holiday Calendar

Crosswalk Labs
Follow this code to learn about emissions data.
Sept 21 - Zero Emissions Day!

June

1 - World Reef Awareness Day
2 - World Peatlands Day
3 - World Bicycle Day
5 - World Environment Day
6 - World Green Roof Day
8 - World Oceans Day
9 - Coral Triangle Day
11 - International Lynx Day

National Oceans Month
National Rivers Month
National Great Outdoors Month

12 - National Cougar Day
15 - Global Wind Day
 - National Electricity Day
16 - World Sea Turtle Day
17 - World Day to Combat Desertification & Drought
20 - World Horseshoe Crab Day
22 - World Rainforest Day
21 - National Up-cycling Day

40

July

1-7th - Clean Beaches Week

Monday of 3rd full week: Coral Reef Awareness Week

3 - Int'l Plastic Bag Free Day
10 - Global Energy Independence Day
12 - Paper Bag Day
14 - Chimpanzee & Orca Day
16 - World Snake Day
26 - International Day for the Conservation of the Mangrove Ecosystems
28 - World Conservation Day
29 - International Tiger Day
31 - World Ranger Day

August
National Water Quality Month

4 - Int'l Clouded Leopard Day
6 - National Tree Day
8 - Int'l Moon Bear Day
10 - Int'l Biodiesel Day
11 - Mountain Day
13 - Int'l Wolf Day
  - World Lizard Day
15 - National Honey Bee Day
19 - World Orangutan Day
26 - African Wild Dog Day
30 - National Beach Day

Last Monday - Friday: National Composits Week

September
National Organic Month

1 - World Beach Day
4 - National Wildlife Day
5 - Amazon Rainforest Day
7 - Int'l Day of Clean Air for Blue Skies
12 - World Dolphin Day
16 - Int'l Day for the Preservation of the Ozone Layer
  - National Clean-up Day

Third weekend in September: Clean-Up the World Weekend

17 - World Manta Day
18 - World Water Monitoring Day
21 - Zero Emissions Day
  - Arbor Day
22 - World Car Free Day

Last Saturday of September: Free Access to National Parks

26 - World Environmental Health Day
27 - National Crush a Can Day

Last Sunday of September: World Rivers Day

29 - Int'l Day of Awareness to Reduce Food Loss and Waste

October
Int'l Walk to School Month

First Monday: World Habitat Day
First Sunday: Change a Light Day

1 - World Vegetarian Day & National Green City Day
3 - World Habitat Day
4 - World Animal Day
7 - National LED Light Day

First Wednesday of October: Energy Efficiency Day

13 - Int'l Day for Natural Disaster Reduction
14 - Int'l E-Waste Day
24 - Int'l Day of Climate Action

Fourth Wednesday: Sustainability Day

November

1 - World Vegan Day
5 - World Tsunami Awareness Day
15 - National Recycling Day
17 - World Fisheries Day

First Friday after Thanksgiving: Green Friday

First Tuesday after Thanksgiving: Giving Tuesday

December

4 - Wildlife Conservation Day
5 - World Soil Day
11 - Int'l Mountain Day
19 - Look for An Evergreen Day

41

Sparks of Thought

Tadpoles try these questions

A - Which chapter had your favorite animal? What would you tell other people about this animal? Why?

B - Name one thing you learned in this book that you could do to help the planet? Why did you choose this?

C - What is one thing that you learned in this book that you didn't know before?

D - What does the blue color mean on the species range maps found on the first page of each chapter?

E - On the species range map, where is the county that you live in?

F - Pick a date from the Environmental Green Holiday Calendar and think of an activity you could do for that day.

G - Which ecosystem do you most want to explore? Why?

H - Have you ever planted a tree? Why is it important to plant native plants?

For a deeper dive try these questions:

AA - If these were dinosaur times, would people be the dinosaurs or the meteor?

BB - Why do pollinators matter?

CC - Why do we need to help the Earth? Are there any plant and animal communities in your area that you can help? How?

DD - Why does the "Shine a Light" poem in the loon chapter differ from all of the other chapters? Why are the stars so important to birds? What does the word "mitigate" mean in that poem?

EE - What is the difference between managed forestry and deforestation? Why does this matter?

FF - How do you discover if there are toxins or other harmful substances in an area or environment?

GG - If an ecosystem gets damaged by human activity, what chain of events could happen?

HH - Are all things humans do harmful? How do we know if we did something that has caused harm?

Resources

Map data and resources for each indicator, respectively:

1. Blanchard's Cricket frog
    1. Blanchard's Cricket Frog (Acris blanchardi) - *Minnesota Amphibian & Reptile Survey. (n.d.). https://mnherps.com/species/acris_blanchardi*
    2. Minnesota Department of Natural Resources - Projects - Frog Survey*: https://dnr.state.mn.us/eco/nongame/projects/cricket-frog-survey.html*
    3. umesc.usgs.gov/terrestrial/amphibians/armi/frog_calls/cricket_frog.mp3
    4. Seek by iNaturalist · *iNaturalist. (n.d.). iNaturalist. https://inaturalist.org/pages/seek_app*
2. Dakota skipper butterfly
    1. Minnesota DNR, Home Nature ETSC Rare Species Guide - Dakota Skipper: *dnr.state.mn.us/rsg/profile.html?action=elementDetail&selectedElement=IILEP65140*
    2. Creating a butterfly garden. (n.d.). *UMN Extension. https://extension.umn.edu/landscape-design/creating-butterfly-garden*
    3. Minnesota Department of Natural Resources. (2023, September 21). *Reintroducing the Dakota skipper to Minnesota's Prairie [Video]. YouTube. https://www.youtube.com/watch?v=a54NErn74VI*
    4. Minnesota Zoo - *Conservation, In Minnesota - Pollinator Conservation Initiative: https://mnzoo.org/conservation/minnesota/saving-minnesotas-prairie-butterfly-heritage/*
    5. Gomez, T. (2023, May 23). Minnesota Butterfly Garden Resources. *Monarch Butterfly Garden- Save the Butterflies. https://monarchbutterflygarden.net/butterfly-garden-resources/minnesota-butterfly-garden-resources/*
3. Red headded woodpecker
    1. Minnesota Biological Survey, *Map data collected by Minnesota Biological Survey. Red-headed Woodpecker http://files.dnr.state.mn.us/eco/mcbs/birdmaps/red_headed_woodpecker_map*
    2. Minnesota DNR- Minnesotas Woodpeckers: *https://dnr.state.mn.us/birds/woodpeckers*
    3. Red-headed woodpecker. (n.d.). Audubon. *https://www.audubon.org/field-guide/bird/red-headed-woodpecker*
    4. Bce_Admin_User. (2024, May 13). Cornell Lab of Ornithology—Home. Birds, Cornell Lab of Ornithology. *https://birds.cornell.edu/home*
    5. The search for the red-headed woodpecker | Three Rivers Park District. (n.d.). *https://www.threeriversparks.org/blog/search-red-headed-woodpecker*
    6. ♫ Red-headed woodpecker - *song / call / voice / sound. (n.d.). Bird-sounds.net. https://bird-sounds.net/red-headed-woodpecker/*
4. Walleye
    1. *Minnesota's Big 10 walleye lakes - Google My Maps. (n.d.). Google My Maps. https://google.com/maps/d/u/0/viewer?mid=1SAz-UxZMzp5-84FzFUvDldi4FnQ&hl=en_US&ll=46.84625137608965%2C-93.68522014999999&z=7*
    2. Stewart, A. (2005). *The earth moved: On the Remarkable Achievements of Earthworms. Hachette UK.*
    3. Illinois Association of Conservation Districts annual meeting, February 2024: Peggy Anesi and Pam Otto presented: *Out Standing in the Field: Tips & Tricks for Teaching Trailside*
    4. *University of Wisconsin Sea Grant. (2020, June 3). Fish ID | Wisconsin Sea Grant. Wisconsin Sea Grant | University of Wisconsin. https://www.seagrant.wisc.edu/fish-id/*
    5. Minnesota DNR website for the Walleye. *https://www.dnr.state.mn.us/fish/catchandrelease.html*
5. Piping plover
    1. Minnesota DNR -Rare Species Guide: Piping Plover: *https://dnr.state.mn.us/rsg/profile.html?action=elementDetail&selectedElement=ABNNB03070*
    2. Nest Watch - Where Birds Come to Life: *https://nestwatch.org/*
    4. American Bird Conservancy. (2023, January 6). *Bird-Friendly Life - American Bird Conservancy. https://abcbirds.org/get-involved/bird-friendly-life/*
    5. Hinterland Who's Who - Piping plover. (n.d.). *https://hww.ca/en/wildlife/birds/piping-plover.html*
    6. eBird - Discover a new world of birding. (n.d.). *eBird. https://ebird.org/home*
    7. *Charadrius melodus (piping plover). (n.d.). Animal Diversity Web. https://animaldiversity.org/accounts/Charadrius_melodus/#:~:text=The%20Piping%20sPlover%20is%20an,cleaning%20the%20Piping%20Plover%20provides.*

6. Zigzag Darner Dragonfly
    1. Minnesota DNR -Rare Species Guide: ZigZag Darner Dragonfly: *dnr.state.mn.us/rsg/profile.html?action=elementDetail&selectedElement=IIODO14160*
    2. MDS - mndragonfly.org. (n.d.). https://mndragonfly.org/
7. Loon
    1. Eckel, E. (2021, December 28). Minnesota Loons - Biodiversity Research Institute - Innovative Wildlife Science Worldwide | Portland, ME USA. Biodiversity Research Institute - Innovative Wildlife Science Worldwide | Portland, ME USA. https://briwildlife.org/loon-program/minnesota-loons/
    2. Journey North Common Loons. (n.d.). https://journeynorth.org/search/Loon.html
    3. National Loon Center. (2024, April 12). National Loon Center Foundation. National Loon Center Foundation. https://nationallooncenter.org
    4. North, J. (n.d.). Journey North Loon Migration. https://journeynorth.org/tm/loon/Babies_KeepSafe.html
    5. Meredith, S. (2023, June 21). Voyageurs Quarter is launched | U.S. mint. United States Mint. https://www.usmint.gov/news/inside-the-mint/voyageurs-national-park-quarter-launch
8. Western moose
    1. Minnesota DNR, Home Nature ETSC Rare Species Guide - Western Moose: https://www.dnr.state.mn.us/rsg/profile.html?action=elementDetail&selectedElement=AMALC03010
    2. Kallok, M. A. K. (2015, September). Moose in Minnesota: This magazine has followed the story of the shifting fortunes of our monarch of the north woods. https://www.dnr.state.mn.us/mcvmagazine/issues/2015/sep-oct/moose-in-minnesota.html. https://www.dnr.state.mn.us/mcvmagazine/issues/2015/sep-oct/moose-in-minnesota.html
    3. Hassanzadeh, E. (2024, March 6). Minnesota's moose population officially stands at 3,470, down from 8,840 in 2006. CBS News. https://cbsnews.com/ minnesota/news/minnesotas-moose-population-officially-stands-at-3470-down-from-8840-in-2006/#
    4. Carey, J. C. (2014). The deepening mystery of moose decline: Biologists are having a hard time pinning down the cause of a moose decline that imperils the species' survival in several states. NATIONAL WILDLIFE®, OCTOBER–NOVEMBER 2014(Sep 29, 2014). https://nwf.org/Magazines/National-Wildlife/2014/OctNov/Conservation/Moose?_ga=2.212741076.1966555935.1713755309-446479990.1713755309
    5. Palm oil – deforestation for everyday products. (n.d.). Rainforest Rescue. https://www.rainforest-rescue.org/topics/palm-oil
9. Green Holiday Calendar
    1. Homes, E. (2023, July 25). A list of environmental holidays - ecotelligent homes. Ecotelligent Homes. https://www.ecotelligenthomes.com/environmental-holidays/
    2. Wikipedia contributors. (2024, May 27). List of environmental dates. Wikipedia. https://en.wikipedia.org/wiki/List_of_environmental_dates